Moths and Butterflies
(Lepidoptera)
of
Hickling Broad
National Nature Reserve

T. N. D. Peet

Published by
Norfolk and Norwich
Naturalists' Society

Occasional Publication No.4

© TND Peet 1992

ISBN   0 9501130 4 2

First published in 1992 by the Norfolk and Norwich Naturalists' Society

All rights reserved

No part of this publication may be reproduced, stored in a retrieval system, or transmitted, in any form or by any means, electronic, mechanical, photocopying, recording or otherwise, without the prior permission of the publishers.

Printed and bound in Great Britain by Witley Press Ltd., Hunstanton Norfolk.

Front Cover: Elephant Hawkmoth – *J.D. Oxenford*
Back Cover: 5 - spot Burnet Moth – *K.W.K. Palmer*

MOTHS and BUTTERFLIES
(Lepidoptera)
of
HICKLING BROAD NATIONAL NATURE RESERVE

Introduction

Broadland is a very special area within the British Isles, and supports a wealth of rare plants, insects and birds. Hickling Broad, with its associated reed-beds, fens and grazing marshes, is of sufficient ecological importance to be designated as a National Nature Reserve and is managed by the Norfolk Naturalists Trust in conjunction with English Nature. The reserve provides a rich habitat for a significant assembly of lepidoptera.

There was no specific list of Hickling lepidoptera until 1974, and that was deficient in records of microlepidoptera. The present list embraces observation made between 1957 and 1989, new species now being added at no more than one or two per season. Light trap records are mainly from Whiteslea Lodge, other sites having included Wagonhill Plantation, Catfield Dyke and the open grazing marshes. Daytime observations and larval hunting have extended over the whole reserve. Recording by the author has been carried out from March to October but the list includes items submitted by D.J. Agussiz, B. Skinner and J.M. Chalmers-Hunt together with day-to-day sightings by the Wardens, the late Colonel R. Sankey and, more recently, S. Linsell. The classification follows 'A Recorder's Logbook or Label List of British Butterflies and Moths' (1979) and the more recent 'An Indexed List of British Butterflies and Moths' (1986) by J.D. Bradley and D.S. Fletcher.

The Habitat

The climate at Hickling is characterized by cold winters with late springs. Summers are warm and often windy, while the autumns are fine and mild. Snow rarely lies for long although frosts are typically hard and persistent. The water of the broad is slightly brackish due to seepage rather than tidal influence. The River Thurne drains the broad and subsequently joins the Bure which reaches the sea at Great Yarmouth, the total length of the two being around twenty miles. There is a tidal rise and fall of six inches at Potter Heigham but this is not noticeable at Hickling. Around the broad, there are extensive beds of reed and sedge which largely depend on man for their survival. The regular harvesting of the two plants prevents the establishment of alder, sallow and birch and the natural succession to wet woodland or carr. Tree development is also suppressed, indirectly, due to the nearness of the coast. In winter, north-easterly winds sweep across the intervening flat grazing land. They are salt-laden, stunting tree growth.

Pure stands of reed are found where systematic cropping takes place. Adjacent to the open water, the reeds are interspersed with greater and lesser reedmace. Away from the water, in drier areas, other species are present.

From May to high summer, there is a succession of plants starting with marsh valerian, yellow flag and ragged robin, to be followed by hemp agrimony and yellow loosestrife together with milk parsley, purple loosestrife, meadowsweet and common valerian. Marsh walls are lined with large blackberry bushes, some bracken and stunted alder, birch and sallow trees. Where bottom dredgings have been dumped on the banks, the bare soil is soon colonised by willowherb and stinging nettle.

The grazing marshes have a coarse grass flora with various rushes and sedges including the occasional patch of cotton-grass. There are two small well-established oakwoods, the oaks being interspersed with alder and hawthorn.

During the last twenty years, there has been a marked decline in the quality of the aquatic vegetation at Hickling. The rare plants *Najas marina*, the water soldier, *Stratiotes aloides*, and the stoneworts have almost gone and the fish population has declined. The water is murky throughout the summer. There are several causes. First, there is pollution, largely due to farm effluent, rich in phosphates and nitrates, draining into the broad. Stirring of the bottom by motor cruisers and excessive dredging to create navigation channels add to the problem.

Twenty four species of butterfly have been recorded on the reserve and nearly five hundred species of moth. Apart from migrants, there are many local rarities among the moth population which are virtually confined to Broadland and their presence illustrates the importance of Hickling as a National Nature Reserve. Some of the more interesting specialities are described in the following text.

Several factors are responsible for the remarkable aggregation of unusual insects at Hickling. The very size of the reserve is important, any local population fluctuation being compensated by the influx of fresh individuals from another part of the reserve. The variety of the vegetation provides a good range of habitats and the East Anglian climate, resembling more closely that of continental north-western Europe than a temperate island, has facilitated the successful maintenance of many western European species.

Butterflies and Butterfly Watching
Hickling is justifiably famous for the swallowtail butterfly, *Papilio machaon*. The British subspecies, *ssp. brittanicus*, is not only different in appearance from the continental form but flourishes in a more restricted habitat with the milk parsley, *Peucedanum palustre*, as the larval foodplant. Observations at Hickling have shown that the butterfly only deposits eggs where the foodplant is not shaded and where there is water within a few metres. The selected plants of milk parsley are usually higher than the surrounding vegetation with the umbels not yet in flower. Whether sight or smell is the trigger for egg laying is not known but butterflies are on the wing from early June until the end of July. The favourite source of nectar is ragged robin and keen photographers should find a good flowering clump, in full sun, and then wait for the insect to appear. In hot summers, a second brood of butterflies may hatch and be flying until late August. Visitors to the Observation Tower may see the spiralling display flight of the insects high above Wagonhill Plantation.

Hickling Broad National Nature Reserve

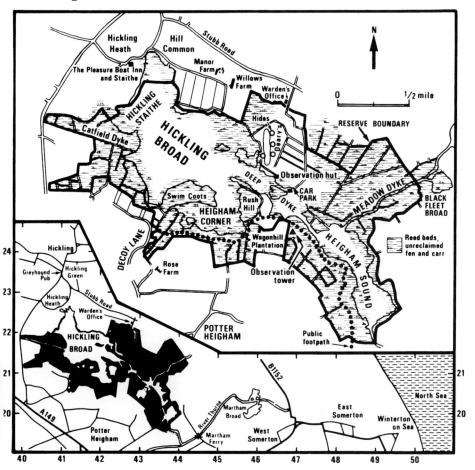

This event has not been recorded in the entomological literature and was first noted in 1969 by the Hickling Warden, Colonel R. Sankey. There is no guarantee of swallowtail sightings even on the sunniest of days but good stations are along Weaver's Way as it passes Wagonhill Plantation and the sleeper track by Deary's Hides. Visitors should remember that the butterfly, whether as an adult or a larva, is protected by law and a photograph taken in the field is the best permanent record of this superb insect.

The small oakwoods around the broad contain fine populations of the purple hairstreak, *Quercusia quercus*. The adult insects fly around the crowns of the mature oaks and are thus difficult to see although they descend to eye level in the late afternoon sunshine. Usually they emerge during the first week in August and the Observation Tower provides an ideal spot from which to observe this butterfly. Any mature oak scrub or wood is worth watching carefully.

Visitors to Hickling will often see superb displays of the peacock butterfly. There are periodic population explosions when the flower-heads of hemp agrimony are crowded with nectar-seeking butterflies. The black spiney larvae are common on nettle and are frequently parasitized.

The first sighting on the reserve of the comma, *Polygonia c-album*, was in October, 1984, to be followed by a second at Catfield Dyke in August, 1987. The species is quite common further inland and may be extending its range.

Although Norfolk is not on a main insect migration route, in most years there are examples of the painted lady, *Vanessa cardui*. Even the spectacular Camberwell beauty, *Nymphalis antiopa*, has been seen, several being noted in the great migration year of 1976.

Low density resident butterflies include the brimstone, *Gonepteryx rhamni*, and the orange tip, *Anthocaris cardamines*, both being spring insects. There is a small colony of the holly blue, *Celastrina argiolus*, at Stubb Mill. Bramble blossom in midsummer is alive with the gatekeeper, *Pyronia tithonus*, and around Wagonhill Plantation is a colony of the shade loving ringlet, *Aphantopus hyperanthus*. Two species of skipper butterfly are present, the large skipper, *Ochodes venata*, and the small skipper, *Thymelicus sylvestris*. Small skipper numbers are usually large but the less widely distributed Essex skipper, *Thymelicus lineola*, does not seem to occur on the reserve.

Moths

The reed leopard, *Phragmatoecia castaneae*, is found only in Broadland. Males are common, coming to light from mid-June until early August, and the slow-growing larvae feed on the rootstocks of the common reed.

The marshland form of the five-spot burnet, *Zygaena trifolii ssp. decreta*, has become very scarce thoughout its range but a small colony survives at Hickling being dependent on the larval foodplant, marsh birdsfoot trefoil, which scrambles over the coarse marshland vegetation.

In 1976, a tiny moth, *Coleophora hydrolapathella*, was discovered breeding at Hickling. The Larvae are very difficult to see, blending perfectly with the foodplant, the great water dock.

A tap on the branches of alders during July will produce good numbers of the scarce *Stathmopoda pedella*, beautifully marked with yellow and black forewings. The larvae of this insect feed on alder catkins.

Among the pyralid group of moths, the rare *Eurryhpara perlucidalis* was discovered in 1974 and apart from Wood Walton Fen this was the only known British location for the species. More recently, it has been found elsewhere in Norfolk and at one or two other suitable sites.

Nascia cilialis and *Evergestis pallidata* are local wetland insects found in rough sedge beds, the former being very scarce nationally.

The bigger moths are more readily noticed by the casual observer. The males of the day-flying emperor, *Saturnia pavonia*, are common in April, and later in the year the large green and black larvae may be seen on bramble.

The oak eggar, *Lasiocampa quercus*, careers at speed over the grazing marshes and along tracks in July and is often mistaken for a butterfly. The drinker, *Philudoria potatoria*, is usually noted as a hairy caterpillar feeding on reed in April and May after hibernation. The adults are common at light in August.

The geometrid group of moths includes a few local specialities. The most notable are a strong population of the rosy wave, *Scopula emutaria*, and two good pug moths. The valerian pug, *Eupithecia valerianata*, is most readily found as a larva, well camouflaged on the ripening fruits of valerian. The dentated pug, *Anticollix sparsata*, comes to light and is common as a larva on yellow loosestrife.

Nine species of hawkmoth have been recorded including such occasional vagrants as the convolvulus, *Herse convolvuli*, and the humming bird, *Macroglossum stellatarum*. Of the resident species the most unusual is the small elephant, *Deilephila procellus*. The lime, *Mimas tiliae*, appears occasionally, wandering from further inland.

Adult males of the puss moth, *Cerura vinula*, come to light but the larvae are difficult to find. The alder kitten, *Furcula bicuspis*, is much scarcer, a single specimen being seen in most years. The maple prominent, *Ptilodontella cucullina*, is another annual visitor, hedgerow field maples being the foodplants.

Hawthorn hedges around the broad are often covered with the hairy larvae of the yellowtail, *Euproctis similis*. The scarcer white satin, *Leucoma salicis*, is frequent at light.

Hickling has two very local species of footman moth. The dotted footman, *Pelosia muscerda*, comes to light in late July and Broadland is its sole habitat in the British Isles. The very rare small dotted footman, *Pelosia obtusa*, is a tiny inconspicuous moth, reluctant to fly to light. Apart from a single specimen from Barton, this is the only locality for the species in the British Isles. It is found in old mixed sedge and reed beds of which there are considerable areas on the reserve.

The large group of Noctuid moths are mostly nocturnal and only recorded by rearing wild larvae or by collection from a moth trap. The wainscot group are particularly well represented as most of them feed on reed. The most notable members of this group are the obscure wainscot, *Mythimna obsoleta*, which is very scarce at Hickling, and the flame wainscot, *Senta flammea*, which is fairly common. The latter is associated with large reedbeds and is found in East Anglia with a few smaller colonies along the south coast. The mere wainscot, *Photedes fluxa*, has been taken twice. It is a true fenland speciality, known better in Cambridgeshire, the larva being confined to *Calamagrostis* grasses. Fenn's wainscot, *Photedes brevilinea*, was discovered new to the British Isles at Ranworth over one hundred years ago and it is common at Hickling. Away from Broadland, the only other site for this species is the reedbed area at Walberswick in Suffolk. The cottongrass, *Eriophorum angustifolium,* of the grazing marshes supports the larvae of Haworth's minor, *Celaena haworthii*, more usually a northern moorland species with only isolated colonies in the south and east of England.

The reed stem-boring wainscots are very well represented. The scarcer ones are Webb's wainscot, *Archanara sparganii*, and the rush wainscot, *Archanara algae*. The silky wainscot, *Chilodes maritimus*, which occurs

7

throughout East Anglia has two striking varieties, *var. wismariensis* and *var. bipunctata*, both of which are found at Hickling.

In recent years the gold spot, *Plusia festucae*, has been divided into two species, the second being Lempke's gold spot, *Plusia gracilis*. The latter is much the commoner in marshland and both occur at Hickling.

Finally, the proximity of the coast, with its extensive dune systems at Sea Palling and Waxham, has led to the capture of many wind-blown, rather than migrant, insects. Species noted include the pyrale, *Anerastia lotella*, the very local pygmy footman, *Eilema pygmaeola*, Archer's dart, *Agrostis vestigialis*, and the lyme grass wainscot, *Photedes elymi*. A note on the coastal insects was published in the Transactions of the Norfolk and Norwich Naturalists' Society in 1983.

Hickling is a particularly open broad lacking the large stands of mature trees found at other more sheltered sites. The only study of the lepidoptera of two different broads was published in 1974 and compared Hickling with Wheatfen. The moths notably absent from Hickling included woodland species such as the lobster, *Stauropus fagi*, the black arches, *Lymantria monacha*, and the alder, *Apatele alni*, together with autumnal insects, possibly overlooked, such as the large ranunculus, *Antitype flavicincta*, and the orange sallow, *Tiliacea citrago*.

Moths conspicuous at Hickling and absent at Wheatfen included the dotted footman, *Pelosia muscerda*, the small dotted footman, *Pelosia obtusa*, the water ermine, *Spilosoma urticae*, and Webb's wainscot, *Nonagria sparganii*. Doubtless further research will modify these inequalities but they highlight the influence of the lack of woodland at Hickling together with the more extensive nature of its marsh-dependent specialities.

Swallowtail butterfly – *D. Mower*
Swallowtail caterpillar – *R. Illingworth*

Comma butterfly – *N.S. Carmichael*
Painted Lady butterfly – *R. Illingworth*

LIST OF SPECIES

Bradley &
Fletcher
'log book
number'   Name

14	Ghost Swift, *Hepialus humuli* *
15	Orange Swift, *Hepialus sylvina* – frequent
17	Common Swift, *Hepialus lupulinus* – common
143	*Nematopogon metaxella* – occasional
148	*Nemophora degeerella* – occasional dancing 'clouds' of males
160	Reed Leopard, *Phragmataecia castaneae* – common: even the females fly to light. Peak emergence mid July
161	Leopard Moth, *Zeuzera pyrina* – twice only, one larva also noted from log of ash.
162	Goat Moth, *Cossus cossus* – very occasional
163	Forester, *Adscita statices* – two records only
170	Five-spot Burnet, *Zygaena trifolii decreta* – a local speciality: see introductory section
186	*Psyche casta* – twice noted on the wing, cases common on reed screen fencing
224	*Triaxomera parasitella* – One at M.V. light August 1988
230	*Monopis crocicapitella*
232	*Monopis monachella* – a single record of this rare moth in 1977
237	*Niditinea piercella* – common in buildings, outhouses and at M.V.
246	*Tinea semifulvella* – occasional
282	*Caloptilia elongella* – once only in 1976
286	*Caloptilia alchimiella* – common
288	*Caloptilia stigmatella*
351	*Phyllonorycter lautella*
358	*Phyllonorycter froelichiella*
385	*Anthophila fabriciana* – abundant wherever nettles grow
410	*Argyresthia brockeella*
411	*Argyresthia goedartella*
412	*Argyresthia pygmaeella*
415	*Argyresthia retinella*
424	*Yponomeuta evonymella* – occasional years of plenty
425	*Yponomeuta padella* – occasional
452	*Ypsolopha nemorella* – occasional
453	*Ypsolopha dentella* – from honeysuckle in oakwoods
457	*Ypsolopha lucella* – one in 1984
460	*Ypsolopha parenthesella* – a few in woodland
464	*Plutella xylostella* – occasionally abundant
470	*Orthotaelia sparganella* – strangely scarce or overlooked: two records only

* No assessment of status where insufficient evidence available

500 *Coleophora hydrolapathella* – a Hickling speciality. Adults fly gently at dusk around the foodplant in July: also comes to light.
501 *Coleophora siccifolia*
532 *Coleophora albidella*
555 *Coleophora follicularis* – poorly observed group: no estimate
559 *Coleophora peribenanderi* of frequency possible
560 *Coleophora paripennella*
566 *Coleophora sternipennella*
595 *Elachista biatomella*
630 *Biselachista albidella*
640 *Batia lunaris* – one or two each year
647 *Hofmannophila pseudospretella* – common
648 *Endrosis sarcitrella* – common
649 *Esperia sulphurella* – frequent
658 *Carcina quercana* – in local oakwoods, common
672 *Depressaria pastinacella* – common
687 *Exaeretia allisella* – one only in 1974
689 *Agonopterix ciliella* – larvae abundant rolling milk parsley leaves
695 *Agonopterix alstroemeriana* – larvae and occasional adults taken
701 *Agonopterix ocellana* – common
710 *Agonopterix conterminella* – one only in 1979
713 *Agonopterix angelicella* – very common
714 *Agonopterix yeatiana* – common
726 *Metzneria metzneriella* – frequent, to M.V. and swept
728 *Paltodora cytisella* – one record
737 *Monochroa palustrella*
745 *Monochroa divisella* – common, particularly to M.V.
748 *Ptocheuusa paupella* – occasional
786 *Bryotropha desertella* – only noted once
787 *Bryotropha terrella*
792 *Mirificarma mulinella* – occasional
796 *Aroga velocella*
802a *Gelechia sororculella*
803 *Gelechia muscosella*
809 *Pexicopia malvella* – once only: larvae feed in seedheads of marsh mallow, a plant found at Martham and Horsey
819 *Scrobipalpa costella*
840 *Reuttia subocellea*
847 *Syncopacma taeniolella*
858 *Hypatima rhomboidella*
866 *Brachmia blandella*
867 *Brachmia inornatella*
868 *Brachmia rufescens* – common, to M.V. and at dusk
870 *Oegoconia quadripuncta* – one only in 1978
877 *Stathmopoda pedella* – some years may be beaten in plenty from alder.
883 *Mompha raschkiella* – occasional
886 *Mompha ochraceella* – frequent

893	*Mompha epilobiella* – common: often found hibernating in October
897	*Cosmopterix lienigiella* – larval mines abundant in reed: not difficult to breed
898	*Limnaecia phragmitella* – numbers to M.V.each year
905	*Spuleria / Blastodacna hellerella* – occasional
926	*Phalonidia manniana* – common
936	Cochylimorpha *straminea* – common, bred out from knapweed
937	*Agapeta hamana* – common
938	*Agapeta zoegana* – occasional
945	*Aethes cnicana* – frequent
968	*Cochylis nana* - one only
969	*Pandemis corylana*
970	*Pandemis cerasana*
972	*Pandemis heparana*
976	*Archips oporana*
989	*Aphelia paleana*
993	*Clepsis spectrana* – adult males to M.V., larvae frequent in May on comfrey and hemp agrimony
994	*Clepsis consimilana*
1002	*Lozotaenia forsterana* – occasional
1006	*Epagoge grotiana*
1010	*Ditula angustiorana*
1016	*Cnephasia longana*
1020	*Cnephasia stephensiana*
1024	*Cnephasia incertana*
1032	*Aleimma loeflingiana*
1035	*Croesia bergmanniana* – one in 1975
1038	*Acleris laterana*
1043	*Acleris aspersana*
1048	*Acleris variegana* – occasional
1053	*Acleris hastiana* – frequent
1058	*Acleris lorquiniana* – larvae in purple loosestrife: adults to M.V., including form striana and f. illiginosana.
1063	*Celypha striana* – occasional
1076	*Olethreutes lacunana*
1089	*Apotomis semifasciana* – one only in 1984
1104	*Endothenia quadrimaculana*
1108	*Lobesia abscisana*
1110	*Bactra furfurana*
1111	*Bactra lancealana* – abundant
1119	*Ancylis geminana*
1120	*Ancylis mitterbacheriana*
1133	*Epinotia bilunana*
1134	*Epinotia ramella*
1136	*Epinotia immundana* – easily bred from rolled alder leaves.
1138	*Epinotia nisella*
1147	*Epinotia cruciana*
1151	*Epinotia trigonella*

1154	*Epinotia caprana*
1169	*Gypsonoma dealbana*
1174	*Epiblema cynosbatella*
1175	*Epiblema uddmanniana* – frequent to M.V.
1176	*Epiblema trimaculana*
1178	*Epiblema roborana*
1182	*Epiblema turbidana*
1183	*Epiblema foenella* – occasional
1197	*Eucosma campoliliana*
1200	*Eucosma hohenwartiana*
1201	*Eucosma cana*
1210	*Rhyacionia buoliana* – two in 1979, windblown or migrants
1255	*Cydia succedana*
1260	*Cydia splendana*
1261	*Cydia pomonella*
1290	*Chilo phragmitella* – very common in every patch of reed
1293	*Chrysoteuchia culmella* – abundant
1294	*Crambus pascuella* – frequent
1300	*Crambus pratella* – occasional
1302	*Crambus perlella* – abundant
1303	*Agriphila selasella* – common
1304	*Agriphila straminella* – frequent
1305	*Agriphila tristella* – abundant
1313	*Catoptria pinella* – frequent
1316	*Catoptria falsella* – a few each year
1325	*Platytes alpinella* – one or two each year
1328	*Schoenobius gigantella* – common and variable
1329	*Schoenobius forficella* – common
1331	*Acentria ephemerella* – occasionally common: best found at dusk on the water's surface
1332	*Scoparia subfusca* – common
1333	*Scoparia pyralella* – occasional
1334	*Scoparia ambigualis* – common
1336	*Eudonia pallida* – common
1341	*Eudonia lineola* – occasional
1344	*Eudonia mercurella* – occasional
1345	*Nymphula nymphaeata* – frequent
1348	*Parapoynx stratiotata* – frequent
1350	*Parapoynx stagnata* - frequent
1354	*Cataclysta lamnata* – abundant
1356	*Evergestis forficalis* – common
1358	*Evergestis pallidata* – a few each year
1376	*Eurrhypara hortulata* – common
1378	*Eurrhypara coronata* – occasional
1380	*Eurrhypara perlucidalis* – first noted in 1974: a scarce moth and at Hickling is a notable speciality
1385	*Ebulea crocealis* – one or two each year
1387	*Nascia cilialis* – a local speciality: frequent most years
1388	*Udea lutealis* – swarms most years

1390 *Udea prunalis* – common
1392 *Udea olivalis* – frequent
1395 *Udea ferrugalis* – intermittently abundant
1398 *Nomophila noctuella* – as 1395: depends on migration
1405 *Pleuroptya ruralis* – larvae plentiful in every nettlebed
1413 *Hypsopygia costalis* – in sheds, reedstacks, and to M.V.
1415 *Orthopygia glaucinalis* – five records only
1417 *Pyralis farinalis* – one or two each year
1428 *Aphomia sociella* – one or two most years
1432 *Anerastia lotella* – three records only: probably windblown from the coast
1433 *Cryptoblabes bistriga* – occasional
1438 *Numonia suavella* – common
1439 *Numonia advenella* – common
1442 *Numonia pempelia* – once only
1445 *Numosia formosa* – occasional
1451 *Pyla fusca* – first noted 1978, occasional
1452 *Phycita roborella* – common
1458 *Myelois cribrella* – occasional
1467 *Ancylosis oblitella* – many in 1976, not seen since
1483 *Phycitodes binaevella* – occasional
1500 *Platyptilia calodactyla* – one only in 1980
1501 *Platyptilia gonodactyla* – common
1510 *Pterophorus tridactyla* – one only in 1983
1517 *Adaina microdactyla* – common
1524 *Emmelina monodactyla* – common
1526 Small Skipper, *Thymelicus sylvestris* – common in grassy places all round the broad. Close examination shows no specimens of the Essex Skipper, *T. sylvestris*.
1531 Large Skipper, *Ochlodes venata* – common in grassy places
1539 Swallowtail, *Papilio machaon* – continues to be frequent all over the reserve. See specific comments in the introduction. Of many larvae bred, no parasites have been seen. House sparrows eat the large larvae on rigid stems of Peucedanum. No recent melanistic sightings, though doubtless they occur.
1546 Brimstone, *Gonepteryx rhamni* – resident: not as frequently as on more sheltered inland broads with more cover, such as Upton.
1549 Large White, *Pieris brassicae* – the least frequent of the Whites
1550 Small White, *Pieris rapae* – common
1551 Green-veined White, *Pieris napi* – common
1553 Orange Tip, *Anthocharis cardamines* – frequent
1557 Purple Hairstreak, *Quercusia quercus* – strong colonies in Wagonhill Plantation and Skoyles, in years of plenty using solitary oaks as well as woods. Best observed from the Observation Tower
1561 Small Copper, *Lycaena phlaeas* – common along dyke walls, often three broods in one season
1574 Common Blue, *Polyommatus icarus* – infrequent
1580 Holly Blue, *Celastrina argiolus* – a few along hedgerows on the

13

Reserve boundary, and a colony at Stubb Mill.
1584 White Admiral, *Ladoga camilla*. Two sightings in Wagonhill Plantation by Stewart Linsell. Frequent in more sheltered broadland woods.
1590 Red Admiral, *Vanessa atalanta* – occasionally abundant, with larvae on nettles
1591 Painted Lady, *Cynthia cardui* –frequency depends on early summer migrants, with locally bred adults in August. Recent "good" years 1982, 1983, 1988, 1992
1593 Small Tortoiseshell, *Aglais urticae* – very common, particularly as nests of larvae on nettles
1597 Peacock, *Inachis io* – abundant, with adults crowding to flowering heads of hemp agrimony in late July
1598 Comma, *Polygonia c-album* – one seen on Skoyles 10 October 1984, a second record Catfield Dyke August 1987
1601 Pearl Bordered Fritillary, *Boloria euphrosyne* – one record only several years ago. Nearest colony was at Sea Palling
1615 Wall, *Lasiommata megera* – frequent along tracks and dry marsh walls. Associated at Hickling with adders
1625 Gatekeeper, *Pyronia tithonus* – common
1626 Meadow Brown, *Maniola jurtina* – common
1627 Small Heath, *Coenonympha pamphilus* – common
1629 Ringlet, *Aphantopus hyperantus* – margins of Wagonhill Plantation, by the Warden's House and the edge of Skoyle's Marsh
1631 December Moth, *Poecilocampa populi* – a few to M.V. in mid-October
1634 Lackey, *Malacosoma neustria* – frequent to M.V. and larvae on hawthorn
1637 Oak Eggar, *Lasiocampa quercus quercus* – females to M.V., males career wildly over the marshes by day in July and early August
1638 Fox Moth, *Macrothylacia rubi* – a few each year: no larvae found as yet.
1640 Drinker, *Philudoria potatoria* – abundant as adults to M.V., and post hibernation larvae common on reed in April
1642 Lappet, *Gastropacha quercifolia* – occasional, never more than one each year.
1643 Emperor, *Saturnia pavonia* – larvae frequent on bramble, adult females occasional to M.V. in May.
1645 Scalloped Hook-tip, *Falcaria lacertinaria* – frequent to M.V.
1646 Oak Hook-tip, *Drepana binaria* – frequent to M.V., larva also found on alder
1648 Pebble Hook-tip, *Drepana falcataria* – common to M.V.
1651 Chinese Character, *Cilix glaucata* – common
1652 Peach Blossom, *Thyatira batis* – frequent
1653 Buff Arches, *Habrosyne pyritoides* – occasional
1654 Figure of Eighty, *Tethea ocularis* – occasional
1657 Common Lutestring, *Ochropacha duplaris* – occasional: always the melanic form

1659	Yellow Horned, *Achlya flavicornis* – frequent in springtime
1663	March Moth, *Alsophila aescularia* – frequent
1665	*Pseudoterpna pruinata* – frequent
1666	Large Emerald, *Geometra papilionaria* – specific to birch: a few each year
1669	Common Emerald, *Hemithea aestivaria* – frequent
1674	Little Emerald, *Jodis lactearia* – occasional
1682	Blood-vein, *Timandra griseata* – common
1690	Small Blood-vein, *Scopula imitaria* – common
1691	Rosy Wave, *Scopula emutaria* – a Broadland speciality: frequent at dusk and to M.V.
1702	Small Fan-footed Wave, *Idaea biselata* – common
1707	Small Dusky Wave, *Idaea seriata* – common
1708	Single Dotted Wave, *Idaea dimidiata* – common
1712	Small Scallop, *Idaea emarginata* – frequent
1713	Riband Wave, *Idaea aversata* – frequent
1719	Oblique Striped, *Orthonama vittata* – occasional, three records only
1724	Red Twin-spot Carpet, *Xanthorhoe spadicearia* – occasional
1725	Dark-barred Twin-spot Carpet, *Xanthorhoe ferrugata* –common
1726	Large Twin–spot Carpet, *Xanthorhoe quadrifasiata* – once in 1981
1727	Silver-ground Carpet, *Xanthorhoe montanata* – common
1728	Garden Carpet, *Xanthorhoe fluctuata* – common
1734	July Belle, *Scotopteryx luridata plumbaria* – once only
1738	Common Carpet, *Epirrhoe alternata* – common
1742	Yellow Shell, *Camptogramma bilineata bilinea* – abundant
1746	Shoulder Stripe, *Anticlea badiata* – frequent in late spring
1748	Beautiful Carpet, *Mesoleuca albicillata* – recorded once July 1980
1749	Dark Spinach, *Pelurga comitata* – occasional
1755	Chevron, *Eulithis testata* –
1757	Spinach, *Eulithis mellinata* – frequent
1758	Barred Straw, *Eulithis pyraliata* – frequent
1764	Common Marbled Carpet, *Chloroclysta truncata* – common
1765	Barred Yellow, *Cidaria fulvata* – occasional
1766	Blue-bordered Carpet, *Plemyria rubiginata ssp. rubiginata* – occasional
1768	Grey Pine Carpet, *Thera obeliscata* – frequent
1776	Green Carpet, *Colostygia pectinataria* – common
1777	July Highflyer, *Hydriomena furcata* – abundant
1789	Scallop Shell, *Rheumaptera undulata* – occasional: three records only, including one from Hundred Acres Wood
1795	November Moth, *Epirrita dilutata* – common
1808	Sandy Carpet, *Perizoma flavofasciata* – frequent
1809	Twin-spot Carpet, *Perizoma didymata* – one in July 1988
1811	Slender Pug, *Eupithecia tenuiata* – occasional to light
1821	Valerian Pug, *Eupithecia valerianata* – seeding valerian heads sometimes alive with larvae
1825	Lime-speck Pug, *Eupithecia centaureata* – common

1830 Wormwood Pug, *Eupithecia absinthiata* – frequent
1834 Common Pug, *Eupithecia vulgata* – occasional
1835 White-spotted Pug, *Eupithecia tripunctaria* – a few
1838 Tawny-speckled Pug, *Eupithecia icterata* – common
1840 Shaded Pug, *Eupithecia subumbrata* – occasional: one or two each year
1842 Plain Pug, *Eupithecia simpliciata* – occasional
1858 V-Pug, *Chloroclystis v-ata* – common
1860 Green Pug, *Chloroclystis rectangulata* – common
1862 Double-striped Pug, *Gymnoscelis rufifasciata* – common
1863 Dentated Pug, *Anticollix sparsata* – adults to light, larvae on yellow loosestrife
1867 Treble Bar, *Aplocera plagiata* – frequent
1874 Dingy Shell, *Euchoeca nebulata* – one record to M.V. 29 July 1984
1882 Small Seraphim, *Pterapherapteryx sexalata* – frequent
1884 Magpie Moth, *Abraxas grossulariata* – frequent
1885 Clouded Magpie, *Abraxas sylvata* – one record to M.V. 1 August 1979, further two in July 1988. Hedgerow elm has virtually vanished from this area of Norfolk
1887 Clouded Border, *Lomaspilis marginata* – abundant, often fluttering on dyke banks at dusk
1889 Peacock Moth, *Semiothisa notata* – frequent
1890 Sharp-angled Peacock, *Semiothisa alternaria* – frequent
1893 Tawny-barred Angle, *Semiothisa liturata* – frequent
1894 Latticed Heath, *Semiothisa clathrata* – occasional
1897 V-Moth, *Semiothisa wauaria* – occasional
1902 Brown Silver-line, *Petrophora chlorosata* – frequent
1906 Brimstone Moth, *Opisthograptis luteolata* – abundant
1907 Bordered Beauty, *Epione repandaria* – common in late August
1908 Dark Bordered Beauty, *Epione paralellaria* – one only in 1960
1910 Lilac Beauty *Apeira syringaria* – single record in 1980
1913 Canary-shouldered Thorn, *Ennomos alniaria* – common
1914 Dusky Thorn, *Ennomos fuscantaria* – common
1915 September Thorn, *Ennomos erosaria* – common
1917 Early Thorn, *Selenia dentaria* – a few
1919 Purple Thorn, *Selenia tetralunaria* – late summer, 2nd brood, adults frequent to M.V.
1920 Scalloped Hazel, *Odontopera bidentata* – occasional
1921 Scalloped Oak, *Crocallis elinguaria* – frequent
1922 Swallow-tailed Moth, *Ourapteryx sambucaria* – frequent
1923 Feathered Thorn, *Colotois pennaria* – frequent
1927 Brindled Beauty, *Lycia hirtaria* – a few in spring
1930 Oak Beauty, *Biston strataria* – a few in spring
1931 Peppered Moth, *Biston betularia* – common. Black carbonaria frequency is about 40%
1937 Willow Beauty, *Peribatodes rhomboidaria* – common
1941 Mottled Beauty, *Alcis repandata* – common
1947 Engrailed, *Ectropis bistortata* – common
1951 Grey Birch, *Aethalura punctulata* – occasional

Emperor moth, male – *K.W.K. Palmer*
Oak Eggar moth, male – *K.W.K. Palmer*

Eyed Hawkmoth, male – *K.W.K. Palmer*
Left: Elephant Hawkmoth caterpillar – *K. Durrant*
Right: Drinker Moth caterpillar – *K. Durrant*

1954 Bordered White, Pine Looper, *Bupalus piniaria* – rare: 3 records only, presumably wind-blown
1955 Common White Wave, *Cabera pusaria* – common
1956 Common Wave, *Cabera exanthemata* – abundant
1957 White Pinion Spotted, *Lomographa bimaculata* – common
1961 Light Emerald, *Campaea margaritata* – frequent
1972 Convolvulus Hawk, *Agrius convolvuli* – two to M.V. August 1984
1976 Privet Hawk, *Sphinx ligustri* – a few each year
1979 Lime Hawk, *Mimas tiliae* – occasional: presumably wind-blown from inland.
1980 Eyed Hawk, *Smerinthus ocellata* – abundant
1981 Poplar Hawk, *Laothoe populi* – common
1984 Humming-bird Hawk, *Macroglossum stellatarum* – one by day August 1986
1990 Striped Hawk, *Hyles lineata* – one to M.V. August 1977
1991 Elephant Hawk, *Deilephila elpenor* – abundant: full grown larvae may be seen on tracks, wandering prior to pupation
1992 Small Elephant Hawk, *Deilephila porcellus* – a few each year to M.V. in June
1994 Buff-tip, *Phalera bucephala* – common
1995 Puss Moth, *Cerura vinula* – adults to M.V. in May: bred through from larvae on sallow
1996 Alder Kitten, *Furcula bicuspis* – one or two examples each year
1997 Sallow Kitten, *Furcula furcula* – common
2000 Iron Prominent, *Notodonta dromedarius* – common
2003 Pebble Prominent, *Eligmodonta ziczac* – very common
2006 Lesser Swallow Prominent, *Pheosia gnoma* – common
2007 Swallow Prominent, *Pheosia tremula* – common
2008 Coxcomb Prominent, *Ptilodon capucina* – common: bred through from larva on alder
2009 Maple Prominent, *Ptilodontella cucullina* – occasional: maple grows locally as a hedgerow tree
2011 Pale Prominent, *Pterostoma palpina* – frequent
2014 Marbled Brown, *Drymonia dodonaea* – occasional
2015 Lunar Marbled Brown, *Drymonia ruficornis* – occasional
2017 Small Chocolate Tip, *Clostera pigra* – once only
2019 Chocolate Tip, *Clostera curtula* – a few each year
2020 Figure of Eighty, *Diloba caeruleocephala* – occasional
2026 Vapourer, *Orgyia antiqua* – day-flying males most years, larvae on nettle
2028 Pale Tussock, *Calliteara pudibunda* – occasional
2030 Yellow-tail, *Euproctis similis* – very common, larvae on hawthorn
2031 White Satin, *Leucoma salicis* – a few to M.V., occasional larvae on reed
2035 Round-winged Muslin Footman, *Thumatha senex* – very common

2037 Rosy Footman, *Miltochrista miniata* – a few
2040 Four Dotted Footman, *Cybosia mesomella* – one record 1980
2041 Dotted Footman *Pelosia muscerda* – locally common: emergence third week July until second week in August
2042 Small Dotted Footman *Pelosia obtusa* – a Hickling speciality since discovered here by C.J. Cadbury in 1963. Optimum emergence last week in July, comes to M.V. reluctantly, a weak flyer.
2044 Dingy Footman *Eilema griseola* – very common
2046 Pygmy Footman *Eilema pygmaeola* – occasional, wind blown from nearby coast. Resident on the Waxham to Palling dune complex
2047 Scarce Footman *Eilema complana* – scarce compared with 2050
2050 Common Footman *Eilema lurideola* – common
2057 Garden Tiger, *Arctia caja* – abundant and variable as larvae and adults. Polyphagous: comfrey and reeds are preferred food plants
2060 White Ermine, *Spilosoma lubricipeda* – common
2061 Buff Ermine, *Spilosoma luteua* – common
2062 Water Ermine, *Spilosoma urticae* – common
2063 Muslin Moth, *Diaphora mendica* – common
2064 Ruby Tiger, *Phragmatobia fuliginosa* – common
2069 Cinnabar, *Tyria jacobaeae* – a few each year
2077 Short-cloaked Moth, *Nola cucullatella* – a few to M.V., larvae on sallow
2080 Square-spot Dart, *Euxoa obelisca* – few
2081 White-line Dart, *Euxoa tritici* – few
2082 Garden Dart, *Euxoa nigricans* – frequent
2085 Archer's Dart, *Agrotis vestigialis* – occasional wind blown from nearby coast
2087 Turnip Moth, *Agrotis segetum* – common
2088 Heart and Club, *Agrotis clavis* – once only
2099 Heart and Dart, *Agrotis exclamationis* – abundant
2091 Dark Sword-grass, *Agrotis ipsilon* – common
2092 Shuttle-shaped Dart, *Agrotis puta* – abundant
2098 Flame, *Axylia putris* – abundant
2102 Flame Shoulder, *Ochropleura plecta* – abundant
2105 Dotted Rustic, *Rhyacia simulans* – first noted 1987, seven examples seen, noted again 1989
2107 Large Yellow Underwing, *Noctua pronuba* – sometimes in over-whelming numbers
2109 Lesser Yellow Underwing, *Noctua comes* – common
2110 Broad-bordered Yellow Underwing, *Noctua fimbriata* – frequent
2111 Lesser Broad-bordered Yellow Underwing, *Noctua janthina* – frequent
2112 Least Yellow Underwing, *Noctua interjecta* – frequent
2113 Stout Dart, *Spaelotis ravida* – once only in 1959
2114 Double Dart, *Graphiphora augur* – occasional

2118	True Lover's Knot, *Lycophotia porphyrea* – occasional
2119	Pearly Underwing, *Peridromia saucia* – frequent
2120	Ingrailed Clay, *Diarsia mendica mendica* – common
2122	Purple Clay, *Diarsia brunnea* – frequent
2123	Small Square-spot, *Diarsia rubi* – common
2124	Fen Square-spot, *Diarsia florida* – frequent
2126	Setaceous Hebrew Character, *Xestia c-nigrum* – common
2127	Triple-spotted Clay, *Xestia ditrapezium* – frequent
2128	Double Square-spot, *Xestia triangulum* – frequent
2130	Dotted Clay, *Xestia baja* – frequent
2133	Six-striped Rustic, *Xestia sexstrigata* – common
2134	Square-spot Rustic, *Xestia xanthographa* – abundant
2136	Gothic, *Naenia typica* – occasional
2137	Great Brocade, *Eurois occulta* – migrant: seen in 1978, 1981, 1982
2139	Red Chestnut, *Cerastis rubricosa* – a few in spring
2145	Nutmeg, *Discestra trifolii* – common
2148	Pale Shining Brown, *Polia bombycina* – occasional
2153	Bordered Gothic, *Heliophobus reticulata* – once only in 1975
2154	Cabbage Moth, *Mamestra brassicae* – common
2155	Dot Moth, *Melanchra persicariae* – common
2157	Light Brocade, *Lacanobia w-latinum* – frequent
2158	Pale-shouldered Brocade, *Lacanobia thalassina* – frequent
2159	Dog's Tooth, *Lacanobia suasa* – a few each year
2160	Bright-line Brown-eye, *Lacanobia oleracea* – abundant
2163	Broom Moth, *Ceramica pisi* – common
2164	Broad-barred White, *Hecatera bicolorata* – occasional
2166	Campion, *Hadena rivularis* – frequent
2173	Lychnis, *Hadena bicruris* – frequent
2176	Antler, *Cerapteryx graminis* – common every year: occasionally abundant
2177	Hedge Rustic, *Tholera cespitis* – frequent
2179	Pine Beauty, *Panolis flammea* – occasional: a known migrant
2182	Small Quaker, *Orthosia cruda* – a few
2187	Common Quaker, *Orthosia stabilis* – common
2188	Clouded Drab, *Orthosia incerta* – common
2190	Hebrew Character, *Orthosia gothica* – abundant
2192	Brown-line Bright-eye, *Mythimna conigera* – frequent
2193	Clay, *Mythimna ferrago* – common
2196	Striped Wainscot, *Mythimna pudorina* – common
2197	Southern Wainscot, *Mythimna straminea* – common
2198	Smoky Wainscot, *Mythimna impura* – common
2199	Common Wainscot, *Mythimna pallens* – common
2201	Shore Wainscot, *Mythimna litoralis* – wind blown from the nearby coast, where this species thrives

2204	Obscure Wainscot, *Mythimna obsoleta* – occasional: two records only 1973,1977
2209	Flame Wainscot, *Senta flammea* – a Hickling speciality: common to M.V. in mid June.
2214	Chamomile Shark, *Cucullia chamomillae* – occasional
2216	Shark, *Cucullia umbratica* – occasional
2217	Star-wort, *Cucullia asteris* – one record in 1979, wind blown presumably from nearby coast.
2225	Minor Shoulder-knot, *Brachylomia viminalis* – first noted 1980
2233	Golden-rod Brindle, *Lithomoia solidaginis* – one in 1976, migrant
2237	Grey Shoulder-knot, *Lithophane ornitopus* – occasional
2243	Early Grey, *Xylocampa areola* – a few
2245	Green-brindled Crescent, *Allophyes oxyacanthae* – occasional
2247	Merveille du Jour, *Dichonia aprilina* – occasional
2250	Dark Brocade, *Blepharita adusta* – twice only
2262	Brick, *Agrochola circellaris* – occasional in autumn
2263	Red-line Quaker, *Agrochola lota* – frequent in autumn
2264	Yellow-line Quaker, *Agrochola macilenta* – frequent in autumn
2265	Flounced Chestnut, *Agrochola helvola* – frequent in autumn
2266	Brown-spot Pinion, *Agrochola litura* – frequent in autumn
2269	Centre-barred Sallow, *Atethmia centrago* – once only
2270	Lunar Underwing, *Omphaloscelis lunosa* – common
2273	Pink-barred Sallow, *Xanthia togata* – frequent
2274	Sallow, *Xanthia icteritia* – frequent: also bred from larvae from sallow catkins
2278	Poplar Grey, *Acronicta megacephala* – common: no melanic forms seen
2280	Miller, *Acronicta leporina* – a few each year: bred from larvae on alder
2283	Dark Dagger, *Acronicta tridens* – common
2284	Grey Dagger, *Acronicta psi* – common
2289	Knot Grass, *Acronicta rumicis* – common
2290	Reed Dagger, *Simyra albovenosa* – two broods each year: larvae found on reed, comfrey, great water dock
2293	Marbled Beauty, *Cryphia domestica* – frequent
2297	Copper Underwing, *Amphipyra pyramidea* – frequent
2299	Mouse Moth, *Amphipyra tragopoginis* – in outhouses, bird hides, sheds etc. and to light.
2300	Old Lady, *Mormo maura* – occasional only: never more than one per year.
2301	Bird's Wing, *Dypterygia scabriuscula* – occasional
2302	Brown Rustic, *Rusina ferruginea* – a few most years
2303	Straw Underwing, *Thalpophila matura* – frequent
2305	Small Angle Shades, *Euplexia lucipara* – frequent
2306	Angle Shades, *Phlogophora meticulosa* – occasionally abundant

2312 Olive, *Ipimorpha subtusa* – two records only
2313 Angle-striped Sallow, *Enargia paleacea* – once only in 1963, presumably a migrant
2314 Dingy Shears, *Enargia ypsillon* – a few each year
2318 Dun-bar, *Cosmia trapezina* – common
2319 Lunar-spotted Pinion, *Cosmia pyralina* – occasional: noted in 1979, 1980
2321 Dark Arches, *Apamea monoglypha* – abundant
2322 Light Arches, *Apamea lithoxylaea* – common
2325 Crescent Striped, *Apamea oblonga* – twice only, in 1958 and 1960
2326 Clouded-bordered Brindle, *Apamea crenata* – frequent
2330 Dusky Brocade, *Apamea remissa* – frequent
2331 Small Clouded Brindle, *Apamea unanimis* – frequent
2334 Rustic Shoulder-knot, *Apamea sordens* – frequent
2336 Double Lobed, *Apamea ophiogramma* – occasional
2337 Marbled Minor, *Oligia strigilis* – common
2339 Tawny Marbled Minor, *Oligia latruncula* – uncommon
2340 Middle-barred Minor, *Oligia fasciuncula* – common
2341 Cloaked Minor, *Mesoligia furuncula* – common
2342 Rosy Minor, *Mesoligia literosa* – a few
2343 Common Rustic, *Mesapamea secalis* – abundant
2345 Small Dotted Buff, *Photedes minima* – common
2348 Lyme Grass, *Photedes elymi* – most years wind blown from the coast
2349 Mere Wainscot, *Photedes fluxa* – three records only: Ammacalamagrostis is the probable food plant
2350 Small Wainscot, *Photedes pygmina* – common
2351 Fenn's Wainscot, *Photedes brevilinea* – common, including a few of the form *sinelinea*
2352 Dusky Sallow, *Eremobia ochroleuca* – a few, common nearby on the dunes
2353 Flounced Rustic, *Luperina testacea* – common
2360 Ear Moth, *Amphipoea oculea* – common
2361 Rosy Rustic, *Hydraecia micacea* – uncommon
2364 Frosted Orange, *Gortyna flavago* – rare
2367 Haworth's Minor, *Celaena haworthii* – a local speciality, larvae on Eriophorum, cottongrass, on the grazing marshes
2368 Crescent, *Celaena leucostigma* – common and variable
2369 Bulrush Wainscot, *Nonagria typhae* – common: pupae in lesser reedmace
2370 Twin-spotted Wainscot, *Archanara geminipuncta* – common: sometimes abundant as pupae
2371 Brown-veined Wainscot, *Archanara dissoluta* – common
2373 Webb's Wainscot, *Archanara sparganii* – a few each year: pupae not yet found

2374 Rush Wainscot, *Archanara algae* – one record only
2375 Large Wainscot, *Rhizedra lutosa* – common and variable in size and markings
2377 Fen Wainscot, *Arenostola phragmitidis* – common
2379 Small Rufous, *Coenobia rufa* – common
2380 Treble Lines, *Charanyca trigrammica* – occasional
2381 Uncertain, *Hoplodrina alsines*
2382 Rustic, *Hoplodrina blanda*
2387 Mottled Rustic, *Caradrina morpheus*
2389 Pale Mottled Willow, *Caradrina clavipalpis* – frequent
2391 Silky Wainscot, *Chilodes maritimus* – common, with all its special varieties, var. *wismariensis*, var. *bipunctata*, var. *nigristriata*
2399 Bordered Sallow, *Pyrrhia umbra* – rare: five records only
2410 Marbled White Spot, *Lithacodia pygarga* – one only
2412 Silver Hook, *Eustrotia uncula* – one only
2413 Silver Barred, *Deltote bankiana* – occasional to light and by day
2418 Cream-bordered Green Pea, *Earias clorana* – frequent
2421 Scarce Silver-lines, *Bena prasinana* – occasional, first record 1979
2422 Green Silver-lines, *Pseudoips fagana* – occasional
2434 Burnished Brass, *Diachrysia chrysitis* – common
2436 Dewick's Plusia, *Macdunnoughia confusa* – migrant: one record 1982
2437 Golden Plusia, *Polychrysia moneta* – a few
2439 Gold Spot, *Plusia festucae* – common: unusual variety with confluent gold spangles has been taken
2440 Lempke's Gold Spot, *Plusia putnami* – common
2441 Silver Y, *Autographa gamma* – abundant
2442 Beautiful Golden Y, *Autographa pulchrina* – frequent
2443 Plain Golden Y, *Autographa jota* – frequent
2444 Gold Spangle, *Autographa bractea* – probably migrant: one August 1989
2447 Scarce Silver Y, *Syngrapha interrogationis* – probably migrant: two records 1982 and 1987
2448 Spectacle, *Abrostola triplasia* – a few each year
2451 Clifden Nonpareil, *Catocala fraxini* – once only in the great migration year of 1976
2452 Red Underwing, *Catocala nupta* – frequent
2466 Blackneck, *Lygephila pastinum* – common
2469 Herald, *Scoliopteryx libatrix* – occasional
2473 Beautiful Hook-tip, *Laspeyria flexula* – occasional
2474 Straw Dot, *Rivula sericealis* – common to M.V. and along marsh banks
2477 Snout, *Hypena proboscidalis* – common, by day among nettles
2484 Pinion-streaked Snout, *Schrankia costaestrigalis* – frequent

2489 Fan-foot, *Herminia tarsipennalis* – a few
2491 Shaded Fan-foot, *Herminia tarsicrinalis* – frequent
2492 Small Fan-foot, *Herminia nemoralis* – frequent
2493 Dotted Fan-foot, *Macrochilo cribrumalis* – a wetland speciality: common

NORFOLK & NORWICH NATURALISTS' SOCIETY
Patron: Her Majesty the Queen

The County's senior natural history society. It has for its principal objectives the practical study of natural science, the conservation of wild life, the publication of papers on natural history, especially those related to the county of Norfolk, arranging lectures and meetings and the promotion of active fieldwork. Specialist Groups cover most aspects of the county's flora and fauna.

Annual Subscription Rates:
Junior £3.00
Ordinary £8.00
Family £10.00
'Natterjack'
Affiliation £15.00

Publications:
May: Transactions
August: Bird & Mammal Report
Quarterly newsletter:

Secretary:
Dr A.R. Leech
3 Eccles Road
Holt
Norfolk NR25 6HJ

Membership Secretary:
C. Dack
12 Shipdham Road
Toftwood
Dereham
Norfolk NR19 1JJ